WACO-McLENNAN COU
1717 AUSTIN
WACO TX 76701

MW01531276

KANGAROO RATS

by Martha London

Cody Koala

An Imprint of Pop!
popbooksonline.com

abdobooks.com
Published by Pop!, a division of ABDO, PO Box 398166, Minneapolis, Minnesota 55439. Copyright © 2022 by Abdo Consulting Group, Inc. International copyrights reserved in all countries. No part of this book may be reproduced in any form without written permission from the publisher. Cody Koala™ is a trademark and logo of Pop.

Printed in the United States of America, North Mankato, Minnesota.

052021
092021

THIS BOOK CONTAINS RECYCLED MATERIALS

Cover Photo: Tom McHugh/Science Source
Interior Photos: Tom McHugh/Science Source, 1, 5 (bottom left), 6, 14; Anthony Mercieca/Science Source, 5 (top); Shutterstock Images, 5 (bottom right); Jim Zipp/Science Source, 9; iStockphoto, 10–11; Rick & Nora Bowers/Alamy, 13, 19 (top), 20; John Cancalosi/Alamy, 17; Robert Shantz/Alamy, 19 (bottom left); Jeff March/Alamy, 19 (bottom right)

Editor: Aubrey Zalewski
Series Designers: Laura Graphenteen and Colleen McLaren

Library of Congress Control Number: 2020948279
Publisher's Cataloging-in-Publication Data
Names: London, Martha, author.
Title: Kangaroo rats / by Martha London
Description: Minneapolis, Minnesota : Pop!, 2022 | Series: Desert animals | Includes online resources and index.
Identifiers: ISBN 9781532169700 (lib. bdg.) | ISBN 9781098240639 (ebook)
Subjects: LCSH: Kangaroo rats--Juvenile literature. | Rodents--Juvenile literature. | Rodents--Behavior--Juvenile literature. | Nocturnal animals--Juvenile literature. | Desert animals--Juvenile literature.
Classification: DDC 591.754--dc23

Hello! My name is

Cody Koala

Pop open this book and you'll find QR codes like this one, loaded with information, so you can learn even more!

Scan this code* and others like it while you read, or visit the website below to make this book pop.

popbooksonline.com/kangaroo-rats

*Scanning QR codes requires a web-enabled smart device with a QR code reader app and a camera.

Table of Contents

Chapter 1
Little Rodents 4

Chapter 2
Leap Across the Sand. 8

Chapter 3
Life Underground 12

Chapter 4
Living Alone. 18

Making Connections 22
Glossary. 23
Index 24
Online Resources 24

Little Rodents

Kangaroo rats are **rodents**.
They are small animals.
Kangaroo rats have
sand-colored fur. Their tails
are **tufted**.

Watch a video here!

Kangaroo rats live in the deserts of western North America. Kangaroo rats move around at night. Nighttime is cooler and less dry than daytime.

There are 22 kinds of kangaroo rats.

Chapter 2

Leap Across the Sand

Kangaroo rats move mostly using their long back legs. They move on their front legs only when they go short distances. They use their front feet to pick up food.

Learn more here!

Kangaroo rats' back legs are strong. The rats jump to move quickly. Their tails help them balance when they jump. Kangaroo rats can jump to escape from **predators**.

tufted tail

Kangaroo rats can jump nearly the length of a small car!

fur

front feet

back leg

Chapter 3

Life Underground

Kangaroo rats eat seeds, plants, and insects. They get water from their food.

The rats carry food in their cheek pouches. They bring it back to their **burrows**.

Complete an activity here!

13

Burrows protect kangaroo rats from danger. Burrows also keep the rats cool during hot days. Kangaroo rats store seeds in their burrows. They eat the stored food during winter.

Kangaroo rats have good hearing. They listen for **predators** such as snakes and owls. Kangaroo rats race to safety if they hear predators.

Living Alone

Female kangaroo rats give birth one to three times per year. Each **litter** usually has two or three babies. Mother rats keep babies safe in **burrows**.

Babies leave their mothers after one to six months. They live alone. They make their own burrows. Kangaroo rats usually live for two to five years in the wild.

Making Connections

Text-to-Self

Would you like to see a kangaroo rat in real life? Why or why not?

Text-to-Text

Have you read other books about desert animals? How are kangaroo rats like those animals? How are they different?

Text-to-World

Kangaroo rats live in burrows. What is another animal that lives in a burrow?

Glossary

burrow – a hole that an animal digs in the ground for shelter.

litter – a group of young animals born to an adult animal at one time.

predator – an animal that hunts other animals for food.

rodent – a type of animal that has a single pair of strong teeth.

tufted – fluffy in one part.

Index

babies, 18, 21

burrows, 12, 15, 18, 21

food, 8, 12, 15

fur, 4, 11

legs, 8, 10

movement, 7, 8, 10

predators, 10, 16

tails, 4, 10

Online Resources

popbooksonline.com

Thanks for reading this Cody Koala book!

Scan this code* and others like it in this book, or visit the website below to make this book pop!

popbooksonline.com/kangaroo-rats

*Scanning QR codes requires a web-enabled smart device with a QR code reader app and a camera.